Symbolic Logic
and the
Binomial Expansion

Two Math Projects

Richard Forringer

Symbolic Logic and the Binomial Expansion: Two Math Projects
by Richard Forringer

Signalman Publishing   2012
www.signalmanpublishing.com
email: info@signalmanpublishing.com
Kissimmee, Florida

ISBN: 978-1-935991-37-3 (paperback)
           978-1-935991-38-0 (ebook)

Library of Congress Control Number: 2011942940

For information on bulk sales or academic sales, call 1-888-907-4423 or email info@signalmanpublishing.com.

Printed in the United States of America

Signalman
Publishing

If found, please return this workbook to:

Name: _____

CONTENTS

Acknowledgments 7
Introduction 8
Notes to the Teacher 11
Notes to the Student 13
Addendum #1 15
Addendum #2 17
Addendum #3 19

Symbolic Logic: An Independent Study 21
    1   Statements 23
    2   Truth Tables 29
    3   Compound Truth Tables 39
    4   Negations 46
    5   Conditionals 55
    6   Converse, Inverse, and Contrapositive 67
    7   Biconditionals 79
    8   Tautologies 89
    Cumulative Review 93
    Conclusion 101

Binomial Expansion Project 102
    Assignments 107
    Addendum #1 109
    Addendum #2 111
    Addendum #3 113
    Addendum #4 115

Summary, Conclusions, and Connections 117

ACKNOWLEDGMENTS

The author would like to thank Dr. Jennifer Bowen and Dr. Jim Hartman, Math professors at the College of Wooster, Wooster, Ohio, for their assistance in proofreading and revising the original manuscript.

Also, a special thank you to Durham Academy for providing a summer work grant that allowed me to write the curriculum for a Logic Course that eventually became the foundation for the Symbolic Logic Project.

And, a special thank you to my son, Dan Forringer, for his work in proofreading the final manuscript.

Thank you also to John McClure at Signalman Publishing for gently guiding me through the process to allow this project to become a reality!

And finally, thank you to my dear friend, Dr. Mary Bader, for the many hours that we worked together writing our books. Without her support and encouragement, this whole process would have been a very different experience.

INTRODUCTION

This book is divided into two sections corresponding to a project. Each project is designed to supplement an advanced course in mathematics. I have recently retired after forty-one years of teaching. By far, the most creative and fun part of my own teaching experience was the development of these two projects, which I used in my classroom in several different ways, depending upon the nature of the courses that I was teaching, and the needs of the students that year. Usually I assigned the logic project during the first semester and the binomial project the second semester, and expected the students to do them as independent projects. During some years, I would take one day a week out of the normal class during a given grading period and put aside the standard curriculum and teach the project. There were some years when I would do just one of the two projects, but most years my students would complete both projects. Even during the years when I would assign just one of the projects, I would hand the other project out and allow those who wanted to do it for extra credit to work on it without any classroom support.

I have taught both of these projects over twenty-five times and in those years, I would revise the content and the structure of the project. The end result—the one you see here—is significantly different from how it all began. I suppose that there will be future revisions as well, but for now I am comfortable with it in book form and hope that teachers and students who work with these materials will find them rewarding, enriching, and enjoyable! Following this introduction, I have written several notes to the teachers who will be using these materials. I would encourage students to read those notes, too, as they provide some of the rationale why I believe this is worth their time to teach. Following that section, I have written several notes to the students who will be working on these two projects. In like manner, I would encourage the teachers to read the notes to the students as well. I have often wondered why publishers produced "Teachers' Editions" and "Student Editions" of texts. It seems that anything worth telling the teacher would be equally worth telling the students and vice versa! As an algebra teacher, I always offered to give my students the "Teachers' Edition" of the text we were using so that they could look at all the clever notes and suggestions that the author provided to the teacher. It seemed to me if they were good enough for me to read, they must be also useful to the student! My hope is that the teachers and students who use these materials will work cooperatively and share with one another their hopes and concerns and good and bad feelings about the process and any insights they gain from the readings. I sincerely hope that each of you will discover a new way of thinking and looking at your world as a result of participating in this process.

Notes to the Teacher

Let me start out by saying: "Good luck." If you are using these materials for the first time, I assume that you will be teaching materials that you are not all that familiar with. Most math teachers have studied both symbolic logic and the binomial expansion, but neither concept is part of the "standard" curriculum in an algebra course, and thus these concepts will provide some new challenges to most teachers. I have established a personal blog that I use exclusively to reply to questions and/or concerns about the material in this text. Please feel free to contact me directly via e-mail (dick@binomial.forringer.net) or by visiting my website (http://binomial.forringer.net). I believe it is most important to help the students see the connections to standard algebraic concepts and the concepts contained in this workbook. I do this by asking leading questions and by providing problem sets that are designed to encourage students and teachers to consider "outside the box" thinking skills. As you do this, be sure to share with your students your insights and elicit their insights as well.

The first thing you need to do as you consider teaching the materials in this packet is to decide how you will be using them. There are several options. I have written these materials to be done as independent projects. Much will depend on the nature of your school, your teaching style, your relationships with your students, and the support of your administration. Think this over carefully before deciding how to present these materials to your students. Part of this decision must include how to grade the projects and how much weight to give them in your averaging system. If you decide to present them to your students as independent projects, be clear with the students about deadlines, assignments, and expectations about each student working independently. Other options are to teach each concept as a part of your standard curriculum. A third option is to give the projects to the students to do for extra credit. It would not surprise me if some teachers come up with other options for presenting these materials to their students. I believe that the materials contained in these two projects will positively impact your students and will provide both the students and the teachers problems that will challenge the thinking processes and will enhance their perception of what mathematics is all about.

Depending on how you decide to use these materials, there are several optional directions that I would suggest using. For instance, when I give the project as an independent study that is done outside of class activities, I often include the following as part of the directions:

As you work on this project, please use the attached time sheet (see addendum #1) to keep careful track of the time you spend. For each session that you work, you should write the date, your starting time and ending time, and a brief, but specific statement of what you did. As you read each section, highlight or underline the most significant information.

Another comment that I would use is:

This independent project is to be your own work. It is not appropriate to discuss it in any significant manner with anyone other than your teacher. If you find a need for extra help, please see me, not another student or tutor or parent. Your signature below indicates that all work on this project is your own. Also, please ask your parent/guardian to read this page and the *important dates* page, so that they will be aware that you are working on this project. I have included as addendum #2 a sample of the *important dates* page so that you will have a notion as to what time frame I used when assigning this project. Clearly, your time frame will differ depending on how you set up the projects. On the *important dates* page, I make reference to a *project check sheet*. I have included an example of the project check sheet as addendum #3.

Again, this is just a sample of what I used and should be modified according to your own needs. There have been years when I have given out one or the other or both of these projects as either extra credit or to be done as a group activity. If they are done as a group activity, you could include in the directions something like:

You are welcomed to work alone or in groups on this project. If you work with others, please be sure that each student in the group is developing an understanding of the problems you are working on. You are welcomed to consult with me or other teachers or tutors or parents to assist you. There will be a test after everyone has handed in the projects and they have been graded and returned. Your grade for this project will be 10 percent from the materials handed in and 90 percent from the test covering these materials. There will be no surprises on the test. If you have completed the project and understand the problems on the worksheets, I anticipate that you will have no problem with the test. I have created two tests that are extremely similar to one another. The cumulative review at the end of the project is comprehensive. If you are comfortable with the problems on the cumulative review, I believe you will be ready to take the test.

NOTES TO THE STUDENT

Let me start out by saying: "Good luck." The materials in this book are not a part of a "standard" algebra curriculum. Please be assured that there is nothing in these pages that is beyond the scope of students who are prepared for algebra. Like many textbooks, these two projects contain chapters that are, for the most part, divided into two parts: an explanation of the material, and then some problem sets to help you to decide if you do or do not understand the explanation. If there are problems that do not make sense, you have several options, depending upon how your teacher has decided to use these materials. He or she will explain how they want you to proceed if you are having difficulties, and I hope that one of your options will be to contact my website, where you will have the opportunity to leave your questions, which I will respond to. I do hope you see these materials as a real opportunity to stretch your thinking process in a way that will allow you to see mathematics as more than just operations with numbers! Mathematics is, indeed, a way of thinking—a style of reasoning—a process of problem-solving skills. There is no doubt in my mind that successful completion of these two projects will provide a different type of foundation for learning those skills.

I hope you decided to read the "Notes to the Teacher," where you will find some information about the mechanics of how all of this will work. Unlike most other math concepts you have learned or will learn, in this case you have direct access to the author. I have a website with FAQ's and a place for you to get in contact with me with your own questions and/or concerns. I hope you will use that opportunity if the need arises. As I said in the introduction, I have taught both of these projects many times, and I still get excited at the notion that you will have the opportunity to study concepts that really can make a difference in how you think!

ADDENDUM # 1
Logic Project Time Sheet

Name = _____

Date	Beginning time (Rounded to nearest 5 min.)	Ending time (Rounded to nearest 5 min.)	Number of minutes worked this session	Total number of minutes worked thus far	Brief, but **specific**, description of what you did this session

ADDENDUM # 2

Important Dates for the Logic Project

1) Starting date: Hand out projects.

2) Two weeks later: Work on project in class (two or three days).

3) A week or two later: First project check.

4) A week or so later: Second project check. At this stage of the project, you should have chapters 1–4 completed, or nearly completed.

5) Two or three weeks later: Hand in time sheet and first 5 chapters of the project. This will be graded and will count for the first marking period.

6) A week or two later: Third project check.

7) A week or two later: Class workday for project. Focus should be on corrections of chapters 1 – 5, if necessary.

8) Two or three days after class workday: Fourth project check. At this stage of the project, you should have completed the first 8 chapters.

9) A week or two later: Hand in "corrections" of the first 5 chapters of the project. This, too, will be graded and count for the second marking period.

10) Several weeks later: Hand in time sheet and chapters 6 – 10 of the project. This part of the project will also count for the second marking period.

NOTE: I have replaced specific dates with generic time frame dates… I recommend that the teacher who is using this project make up an "Important Dates for the Logic Project" sheet with specific dates included.

ADDENDUM # 3

THIS PAGE IS TO BE COMPLETED AND TURNED IN ON _____.

Name = _____   Date = _____

Please fill in the approximate percent of completeness for each of the following. Your response should end in zero… [0, 10, 20, 30…100 percent]

1) Read, studied, and highlighted chapter 1 _____
2) Completed problem set # 1   _____
3) Read, studied, and highlighted chapter 2 _____
4) Completed problem set # 2   _____
5) Read, studied, and highlighted chapter 3 _____
6) Completed problem set # 3   _____
7) Read, studied, and highlighted chapter 4 _____
8) Completed problem set # 4   _____
9) Read, studied, and highlighted chapter 5 _____
10) Completed problem set # 5   _____
11) Read, studied, and highlighted chapter 6 _____
12) Completed problem set # 6   _____
13) Read, studied, and highlighted chapter 7 _____
14) Completed problem set # 7   _____
15) Read, studied, and highlighted chapter 8 _____
16) Completed problem set # 8   _____
17) Completed the cumulative review.   _____
18) Read, studied, and highlighted the conclusion _____
19) Do you need extra help with this project? _____
20) Questions, Comments, or Concerns? Is there anything you want to
 ask me or tell me relative to your work on this project? _____

_____

_____

SYMBOLIC LOGIC

An Independent Study

This project contains materials that are designed to be learned "on your own" by reading and looking at examples and doing problems. While *symbolic logic* is typically not a topic introduced in algebra, it contains many concepts that are "connected" to a traditional algebra curriculum and does, therefore, enhance the program. Symbolic logic helps us develop the ability to think abstractly—the ability to reason both from direct and indirect information.

This study is divided into 10 chapters. Following most chapters are a series of problems for you to complete. These problems are designed to determine if you understand the material you read. If any of these chapters, or parts of these chapters, do not make sense, then please see your teacher for further explanation.

1) Statements
2) Introduction to the Truth Table
3) Compound statements
      a) Conjunctions
      b) Disjunctions
4) Negations
5) Conditionals
6) Converse, Inverse, and Contrapositive
7) Biconditionals
8) Tautologies
9) Cumulative review
10) And in Conclusion...

CHAPTER 1

Statements

In mathematics, we take *numbers* and perform *operations* on those numbers to solve problems. Similarly, in symbolic logic we take *statements* and perform different types of *operations* on them to solve problems of logic and reasoning. In logic, a statement is defined as, "any meaningful assertion that is definitely true or false." Just like in algebra, we are apt to use letters such as "x" and "y" and others to represent numbers; likewise in logic, we use letters such as "P" and "Q" and others to stand for statements. We are able to make generalizations using symbols to stand for either *numbers* (math) or *statements* (logic). Much of this study will be to learn and understand the symbols used in logic and how to apply them to problem sets.

An expression is not a *statement* if it cannot be said that it is either true **or** false. For instance, "It is hot out" is a sentence that is not a statement. It is an opinion. Any question, phrase, or incomplete thought is not a *statement* in the logistic sense of the word. For a sentence to be a *statement*, it must be a meaningful assertion that is definitely true **or** false.

Any sentence that contains a variable is called an "open sentence" and is therefore not a *statement*, as the value of the variable is unknown. For instance, $x + 4 = 12$ is an example of an "open sentence" and therefore not a *statement*, as it is possibly true or possibly false, depending on the value of "x". In like manner, "She is cute" is an *open sentence*, with the word "she" being the variable.

Some say that mathematicians and logicians are "lazy" people because they are always looking for shortcuts. I prefer to use the word "efficient". But whatever we call it, it cannot be denied that logicians and mathematicians seem to be able to rewrite lots of what they do using symbols rather than words. Indeed, that is one of the fundamental facets of symbolic logic! Instead of saying, "A square is a quadrilateral," logicians will assign a letter to that statement. Thus, we could state that "P" is just a shortcut way of writing, "A square is a quadrilateral." In like manner, the variables used could stand for any statement that we wish to assign to them, and we can reassign new statements to letters (variables) whenever we wish. As we think about that, it is not so unusual, as we do the same thing in algebra all the time. In the sentence:

$$x + 5 = 7,$$

$$x \text{ is } 2.$$

But in the sentence:

$$x + 5 = 10,$$

$$x \text{ is } 5.$$

Note that "x" can be any number we want. Likewise "P" can represent any meaningful assertion we assign to it. So in one problem, "P" can stand for, "A square is a quadrilateral," and in another, "P" could represent "water is a liquid," or any other statement we choose! In like manner, we can do the same for other variables. It is only when the letters are assigned meaning that the *symbols* in *symbolic logic* can be used to solve problems.

Problem Set #1

For each of the problems 1 – 27, determine if the expression is or is not a *statement*. If it is a statement, state if it is true or false. If it is not a statement, state why. Some reasons could be that it is an open sentence, that it is an opinion, that it is a command, that it is a question, or that it is a phrase.

Example # 1 George Foreman is president of the United States.
Example # 2 That car is expensive.
Example # 3 There are 24 hours in a day.

Example #1 = Yes, it is a statement. It is false.
Example #2 = Not a statement: "that car" could refer to any vehicle, and expensive is a relative term.
Example #3 = Yes, it is a statement. It is true.

1) Every square has 4 sides. _____

2) New York City is the capital of North Carolina. _____

3) He is 13 years old. _____

4) Problem number 11 in this set of problems is not a statement.

_____

5) How many prime numbers are divisible by 5? _____

6) Problem number 27 in this problem set is a false statement. _____

_____

7) $7 - 9 = 2$ _____

8) $x^2 = 25$ _____

9) Today is Sunday. _____

10) Problem # 3 in this set of problems is a statement. _____

_____

11) It is false that $7 - 9 = 2$ is a false statement. _____

12) Every statement that is not false is true. _____

13) Every number that is not even is odd. _____

14) Name a number between 10 and 20. _____

15) Tuesday follows Monday. _____

16) Algebra is fun. _____

17) There are 6 natural numbers that will divide into 12. _____

18) So far, this project is easy! _____

19) $12 + 2 \div 2 = 7$ _____

20) There are two one-digit prime numbers. _____

21) No number is both positive and negative. _____

22) Every number is either positive or negative. _____

23) There are 4 natural numbers that will divide into 12. _____

24) There are 7 natural numbers that will divide into 12. _____

25) This sentence contains the letter "e" nine times. _____

26) Twenty is divisible by 2 and 4 and 5 and 20. _____

27) Problem number 1 in this set of problems is a false statement. _____

_____

CHAPTER 2

Truth Tables

Truth Tables are a good method of establishing patterns that indicate all possible conditions that might exist for a given series of statements. Truth tables provide a compact method to study sentences or compound sentences in a systematic manner. They are a powerful tool for changing the form of a problem. The initial truth tables, introduced here, may seem simplistic, as there are so few options. But, as the combinations and numbers of statements increase, truth tables provide a way to organize the information to allow us to visualize facts easier. Since a statement, by definition, must be either true or false, the simplest truth table to express that thought would look like this:

P
T
F

The top row of every truth table is always reserved for the name of the variable(s). Many truth tables will have multiple rows and columns with multiple possibilities to be considered. This, in fact, will be the next notion we will consider. This first table has only one variable: "P". Remember, "P" is a variable that stands for a statement. Since there are only two possibilities for any one statement, i.e., it has to be "true" or "false", there are only two rows in the table itself [remembering that the top row is just the name(s) of the variable(s)]. Let us now consider working with more than one variable. First, what if there are two? I will name the variables "P" and "Q". Now, each variable could be true or false. So, they both could be true, or they both could be false, or the first one true and the second one false or the first one false and the second one true. It is important to ask (and answer) the following question: Are there any other possibilities other than the four just listed? Think about that... The correct answer to the question is "No." By definition, a statement is a meaningful assertion that is either true or false, so if we have two statements, there are only four possibilities and they are compactly summarized in the following two-variable truth table.

P	Q
T	T
T	F
F	T
F	F

The order shown in this table is important for two reasons. First, it minimizes the chances of missing a possibility and second, it allows us to find specific cases easier. There are several methods for remembering the order, but they are easier to explain and understand with a three variable truth table. Let us consider one of those now.

In each of the two truth tables we have just looked at, the top row was reserved for the naming of the variables. Therefore, in the first table there were two rows to give the two possibilities (for one statement), and in the second table there were four rows to give the four possibilities (for the two statements). Do you want to predict how many rows there will be for three or four or five variables? The pattern will (I hope) make sense after I explain it, but sometimes it is more fun to try to figure out the pattern before it is shown to you. Here are some hints... With one variable there were two rows, with two variables there were four rows, and with three variables there will be "x" rows.... Let's list the data to see if it helps us see the pattern.

Number of variables	Number of rows
1	2
2	4
3	x
4	y

So, it seems that the question we need to be asking ourselves is, "How many possibilities are there if there are three variables, then again, if there are four variables?" One way to figure this out would be to list all the possibilities. That could be difficult if we do not have a pattern, as we might skip one or two or more possibilities. Logicians have come up with several methods to assure that we will find all the possibilities, and thus have truth tables that will allow us to consider every option. Let's look at a truth table with three variables. Again (as always) the top row is used just to list the variables. In this case, I am going to call them "P", "Q", and "R."

P	Q	R

Clearly, one row must be all three are true, and another row must be that all three are false. Then there are all those other possibilities... the first two true with the last one false... the first and last one true with the middle one false... the first one true with the other two false... and on and on! So, the big question to be asked now is, "How can we put these in a reasonable and logical pattern to be sure of getting every possibility in an organized way?" If you haven't figured it out, there will be 8 rows. Here is the truth table for three variables, with an explanation as to how it was put together.

P	Q	R	
T	**T**	**T**	Notice that in the column the furthest to the right (in this case, the "R" column) the occurrences of T and F alternate, starting with T. In the next column to the left, (the "Q" column) the T's and F's alternate in pairs, again starting with T. In the next column to the left, (the "P" column) the T's go in groups of 4, and then the F's go in groups of four. Notice that the 8 "T's and F's" that are in bold print are the same as the original truth table with just two variables. That same pattern occurs again, underneath in italicized print.
T	**T**	**F**	
T	**F**	**T**	
T	**F**	**F**	
F	*T*	*T*	
F	*T*	*F*	
F	*F*	*T*	
F	*F*	*F*	

There are actually several different methods of describing what the patterns are with truth tables. One way is to say that the column to the furthest to the right always **alternates** with T and F, starting with "T". Notice that this was true with a single variable truth table and a two variable truth table as well as the three variable truth table. It is also true that the second column over from the right will **alternate in**

groups of two, (again, starting with T) then the next one in groups of 4, then the next one in groups of 8, then the next one in groups of 16, and so on... The number of rows in a given truth table can be expressed as a power of two, with the exponent being the number of statements in the table. Thus, a truth table with one variable would have 2^1 number of rows, or two. When there are two statements the exponent of two is two, when there are three variables, the exponent of two is three, and so on. This can be summarized in the following chart.

Number of variables	Number of rows in the truth table
1	2^1 or 2
2	2^2 or 4
3	2^3 or 8
4	2^4 or 16
5	2^5 or 32

Notice that with 6 variables, there would be 64 rows in the truth table, a number that is almost unimaginable to put together without a *system*. The number of rows grows *exponentially*. Thus, if we were to consider the truth table with 30 variables, there would be 2^{30} or over 1 billion rows! That, of course, means that there are over 1 billion possible combinations. Since each row would contain 30 T's and F's, there would be over 30 billion total letters that would have to be written to make such a truth table! To put that in perspective, if one were to write a T or an F at the rate of 1 per second, it would take approximately 1000 years to set up the problem! If we were to write a truth table with 15 variables, we would need to write (15) 2^{15} total T's and F's! The good news is that there is rarely cause to consider logic problems that use more than three variables. In this project, we will occasionally use a truth table with 5 variables, but for the most part will consider only truth tables with two or three variables! The number of rows and columns of a truth table are a function of what is called a *binary* system. There are many areas of mathematics and science that are based on binary code. Besides symbolic logic, probability is one of the common studies that have a similar binary basis. Consider when one coin is tossed. There are two possibilities: it could land heads (H) or tails (T). Consider when two coins are tossed. They could be two heads or two tails or the first head and the second tail or the first tail and the second head. What about if three coins are tossed? There would be 8 possible outcomes that could be summarized in the following chart. The second project in this text, the binomial expansion, will deal with this problem from a different perspective.

Coin # 1	Coin # 2	Coin # 3
H	H	H
H	H	T
H	T	H
H	T	T
T	H	H
T	H	T
T	T	H
T	T	T

Does this chart look familiar? Compare it to the previous chart with the three variables "P" and "Q" and "R". If we replace each "head" with "T" and each "tail" with "F", the two charts are identical. That must mean that the same set of possibilities exist for the two different problems sets! You will come across this *pattern* of a binary distribution in the second project, and certainly in years to come as you continue your study of mathematics.

Problem Set # 2

1) Write a truth table for 5 variables. What you will be filling in is called the *body* of the truth table.

L	M	N	P	Q

2) How many rows _____ and how many columns _____
would you need to write a truth table for 10 variables? How many (total)
T's and F's would you need to write in the body of the table?

_____

3) "There are always the same number of T's and F's in the body of any
correctly constructed truth table" is or is not a statement. _____

Why? _____

4) "Describe" the last row of the body of a truth table that contains 38
variables. _____

5) "Describe" the next to last row of the body of a truth table that contains
38 variables. _____

6) If you toss 4 coins, how many **different** ways are there to get two heads
and two tails? _____ Draw a table, similar to the "coin #1, coin
#2, coin #3" table previously displayed to show all possibilities.

7) If you constructed the table for problem number 6 correctly, you might be able to discern that there are a total of 5 different potential outcomes when one tosses 4 coins. Those five outcomes are:

    1) They are all heads.
    2) Three of them came up heads, one of them tails.
    3) Two of them came up heads, two of them tails.
    4) One of them came up heads, three of them tails.
    5) They are all tails.

There is only one way for the first of those potential outcomes to happen, and that is represented by the first row of the 4-coin table... There are 4 ways for the second of those possible outcomes to happen. Please complete the following table, which summarizes how many different ways there are for each of the potential outcomes to occur....

Potential outcomes	How many ways that could happen?
They are all heads.	
Three of them came up heads, one of them tails.	
Two of them came up heads, two of them tails.	
One of them came up heads, three of them came up tails.	
They are all tails.	

8) What do you get when you add the 5 numbers you wrote in the chart above? _____ What is the significance/relevance of that number?

_____

9) If you were to toss one coin, how many different outcomes are possible? _____

10) If you were to toss two coins, how many different outcomes
    are possible? _____

11) If you were to toss three coins, how many different outcomes
    are possible? _____

12) If you were to toss four coins, how many different outcomes
    are possible? _____

13) If you were to toss five coins, how many different outcomes
    are possible? _____

14) If you were to toss six coins, how many different outcomes
    are possible? _____

15) If you were to toss "x" coins, how many different outcomes
    are possible? _____

16) According to what you read in this chapter, it would take approximately
    1000 years to complete a truth table with 30 variables. What if you
    added one more variable? Approximately how much longer would it
    take to do a 31 variable table? _____

CHAPTER 3

Compound Truth Tables

Conjunctions

Up to now, truth tables are just mechanical ways of organizing possibilities of combined information about whether or not a series of given statements are true or false. To take it from the mechanical to the useful, we need to add more information or options. Compound sentences are taking a series of statements and connecting them by words. The truth or falseness of compound sentences depends on the truth or falseness of the separate statements and the connective word(s) or symbol(s) used to form the compound statements. The first connective word that we will consider is the word "and". At this point, I must take a short digression! I will come back to "and" statements momentarily... One of my personal "pet peeves" is some of the choices mathematicians have used to pick words or symbols to stand for mathematical terms. For instance, all of the adjectives, "real, rational, irrational, imaginary, odd and natural" are common every day English words that have meaning both inside of and outside of the world of mathematics. Each of those six adjectives is used to describe a set of numbers. Yet, we associate "natural" in the English language with notions such as "something that really exists, or something that is common." Are there "un-natural" numbers? I personally believe that all six of these words are, or can be, confusing to those learning about number systems. Are "natural numbers" more "natural" than "whole numbers" or "rational numbers"? Similar questions can be asked about all of those other words—poorly chosen words, in my opinion! Likewise, the symbols for **subtraction** and **negative** are remarkably similar. They can be, and are, confusing to many students first learning about subtraction as the adding of the opposite! Again, a very poor choice as far as I am concerned! The word "**or**" in our English language has two meanings: the *inclusive* **or** or the *exclusive* **or**. Four different students could correctly and reasonably answer the following question four different ways, and each of them could be correct: "Do you take French or Latin?" One could answer "French", one could answer "Latin", one could answer "no" and a fourth could answer "yes"... all depending on whether they were considering the word "**or**" to be used in its inclusive or exclusive sense. More on this later... I did say, "a short digression", so back to the topic at hand: the word "**and**". When two statements are connected by the word "**and**", it is said to be a **conjunction**. Fortunately, in this case, the word "**and**" in our common everyday English usage and in the more technical symbolic usage, are the same! The word "**and**", used as a connector in symbolic logic, means, as the word implies, "both". So, if I assign the letter "P" to refer to the statement, "Five is a prime number" and I represent "Five is an odd number" with the letter Q, then "P **and** Q" is a true statement because both of the given statements are true. A **conjunction** is true if and only if both of the connected statements are true. Again logicians, being the lazy (efficient) people that they are, have come up with a symbol to replace the word "**and**". That symbol is "∧". Thus "P ∧ Q" would be read, in this case as: "Five is a prime number and five is an odd number." Since five is both a prime number and an odd number, then the symbolic statement "P∧Q" is true. This can be expressed with a truth table in the following manner:

P	Q	P∧Q

First of all, we will fill in the body of the truth table with our "standard" pattern we established in the previous chapter.

P	Q	P∧Q
T	T	
T	F	
F	T	
F	F	

Now, in the top row of the body of the truth table, we have P being true and Q is also true. In the second row Q is false, in the third row P is false and in the last row they are both false. Therefore, it is only in the top row that the conjunction P∧Q is true and our completed truth table would look like this:

P	Q	P∧Q
T	T	T
T	F	F
F	T	F
F	F	F

While it took me three truth tables to explain how this worked, in the future it will take just one, as we will not need to go through the details of how the body is set up each time, nor will we need to create a definition of **conjunction** each time. The **conjunction** is True only when both of the two connected statements are True. The **conjunction** truth table established above is the first of many compound truth tables we will be considering. The truth tables for all compound sentences are a function of the values of the simple sentences used within the compound sentence. A compound truth table is just a truth table that considers two or more statements connected with some sort of connective word. The first of those connective words that we considered was the word "**and**" and the next will be the word "**or**".

Disjunctions

Part A of chapter three was about two statements connected with the word "**and**", which are called **conjunctions**. Part B of chapter three is about two statements connected with the word "**or**", which are called **disjunctions**. As stated in the previous digression, while the word "**and**" typically has one and only one normal meaning, the word "**or**" can be interpreted two significantly different ways in the English language. There is the *inclusive* **or** which means one-or-the-other-or both.... and the *exclusive* **or** which means "just one of the two". Within our normal, everyday usage of the word "**or**", we typically are using the *exclusive* **or**. When I go to a restaurant and order coffee and my waiter asks the question, "Do you want regular or decaf?", I understand him/her to mean which do you want. However, it is possible to interpret that question to imply: "We have two types of coffee: regular and decaf, do you want one of those?" The answer would, of course, be "Yes."

One year... many years ago... I gave a "true or false" quiz to one of my algebra classes. It was a rather lengthy quiz with some difficult questions that I thought would take my students the entire class to finish. Fifteen seconds after handing out the quiz, one of my students returned it with the word "yes" filled in each blank. He, of course, had interpreted my directions as asking if each of the problems was "true or false", just as I had written at the top of the page. So, he asked himself, is problem number one true or false? Yes it is, he concluded, and wrote down the correct answer... "Yes!" What about problem number two? Again, he concluded that it indeed was true or false, so he continued to write the word "yes" for each answer! Consider the following situation: On a given Tuesday morning a friend of yours asks the question: "Is today Wednesday or Thursday?" Think about it.... Might you reasonably and correctly reply with the word, "No", and might your friend feel that your response was reasonable and gave him the information he was requesting? On the other hand, what if the question had been, "Is today Monday or Tuesday?" You would probably reply with the word, "Tuesday", since it is most reasonable to interpret the question as using the *exclusive* **or.** But, might a reasonable reply have been, "yes"? Yes, today is Tuesday or Wednesday.... so, just as my student had replied to the "true or false" question with "yes", so too could you reply to the "Tuesday or Wednesday" question with "yes"! In "normal" conversation or in context, it is usually self-evident which form of the word "**or**" is intended. To make life easier for my students and me, I have mandated that when used with me, the word "or" will always be used in its *inclusive* sense! Thus, when I hand out a paper in class and a student asks me is this a quiz or homework, I will give the appropriate answer of "yes" or "no". As was true for the conjunction situation, logicians also have a symbol for "or", and it looks like this: ∨. Thus, P ∨ Q is read "P or Q" while P ∧ Q is read "P and Q". An "**or**" statement is called a **disjunction** and the truth table is as follows;

P	Q	P∨Q
T	T	T
T	F	T
F	T	T
F	F	F

Note that a **disjunction** is false in only one circumstance, when both connected statements are false, the last row. A **disjunction** that you have already used in mathematics is the symbol "≤". We read the statement $4 \leq 7$ as, "Four is less than or equal to seven." Another way to read $4 \leq 7$ is: "Four is less than seven or four is equal to seven." Since a **disjunction** is true when either one or the other or both of the connected statements is true, $4 \leq 7$ is a true statement!

Problem Set #3

Given that P stands for, "A dime is a coin" and Q stands for, "All coins have value", write in words:

1) The **disjunction** P $\lor$ Q. _____

2) The **conjunction** P $\land$ Q. _____

Construct a truth table for both the **conjunction** and the **disjunction**:

P	Q		

3) Given that R stands for, "All squares have 4 sides," and T stands for, "All squares are quadrilaterals," explain why R $\land$ T is or is not true._____

_____

4) Give a situation, statement or sentence that is different from any listed in this chapter showing a circumstance where the word "**or**" is used in the *inclusive* sense. _____

5) Give a situation, statement or sentence that is different from any listed in this chapter showing a circumstance where the word "**or**" is used in the *exclusive* sense. _____

6) Here is a direct quote from a sentence used in this chapter: "In normal conversation **or** in context, it is usually self-evident which form of the word 'or' is intended." Consider the use of the word "**or**" in that sentence and describe if it was intended in the *inclusive* **or** *exclusive* sense. _____

_____

Note that the word "**or**" is also used in the preceding sentence. Which use of the word "**or**" was indicated in that sentence? _____

7) Are you in 7th grade or 8th grade? _____

8) Is your name Laura or Sarah? _____

9) Are you left handed or right handed? _____

10) What are three potential answers to problem number 9 that you believe might be accepted as correct?_____   _____

_____

11) If your name indeed were Laura, how would you have answered problem number 8 in this set of problems? _____

12) There are two reasonable and correct answers to problem number 11. What are they? _____   _____

13) Give a truth table for the following: [(P ∧ Q) ∨ (P ∨ R)] ∨ Q

P	Q	R	P∧Q	P∨R	(P∧Q)∨(P∨R)	[(P∧Q)∨(P∨R)]∨Q

Endnote: By the way... The student who answered "yes" to every problem on the "true or false" quiz received a 100% for that quiz. I did tell him, and the rest of the class, that I gave him credit for his clever and gutsy response to my directions, but included that I would only do this once, honoring his being the first to come up with such an idea! I told him that I would neither give credit again for a student doing this, nor would I expect any other teacher to do so!

CHAPTER 4

Negations

Conjunctions and **disjunctions** require two simple sentences to be joined by a connecting word or symbol that represents a word. A *binary operation* in mathematics is an operation that requires two numbers. For instance, addition is binary. So is multiplication. We need two numbers to add or multiply. An example of a mathematical operation that is not binary is, "taking the square root of." *Taking the square root of* is an operation that is performed on ONE *not* TWO numbers. So, for me to say something like, "What is the square root of 49?" would be a meaningful question. However, for me to say, "What do you get when you add 7 with?" is not meaningful, as addition requires two numbers! In symbolic logic, most of the symbols we use require two statements that are joined by a symbol. We have already discussed conjunctions and disjunctions and will soon be discussing several more. But, there is one symbol that is used with one and only one statement, and that is the symbol for *negation* or the opposite of. That symbol is "~".

Thus, if P represented "Today is Sunday," then ~P would be read, "Today is not Sunday." The negation of a statement will always have the opposite truth-value as the statement itself. Thus, whenever "P" is true then "~P" is false, and whenever "P" is false then "~P" is true. For instance:

Symbol	Statement in words	Truth value
P	There are 7 days in a week.	True
~P	There are not 7 days in a week.	False
Q	$5 + 7 = 10$	False
~Q	$5 + 7 \neq 10$	True

The negation of a true statement is always false, and the negation of a false statement is always true. Note that there are often several variations of ways to write the negation of a statement. For instance:

 P: A right angle contains 90 degrees.

~P: A right angle does not contain 90 degrees.

~P: It is not true that a right angle contains 90 degrees.

Truth tables containing negations and conjunctions and disjunctions can get a little more complicated than those in the preceding chapter. Some of these will appear in the following problems. "Building" a truth table is a three part process. First, we must do the headings. The headings are the top row of the table, and are dependent on the problem to be solved. For instance, suppose we were doing a truth table for ~ [P ∧ ~(Q ∧ ~R)]. We note that there are three variables, so our first three entries in the truth table will be the three variables. The "starting point" for the truth table follows. The last entry in the heading's row will always be the problem itself.

P	Q	R							~ [P ∧ ~(Q ∧ ~R)]

The number of entries needed in between the three basic variables and the last heading entry will vary dependent on the problem itself. Just like there are rules for order of operations in algebra, there are rules for the order we do truth tables. Parentheses and brackets are the key... We always start with the inner most symbols of inclusion. In this case, we start with Q ∧~R, however, before we can do ~R, we must have the R

column, which, of course, we do have in the third heading. Once we have done the
Q ∧~R column, we can negate that and do the ~(Q ∧~R). We then do the conjunction
of P with ~(Q ∧~R). Once we have completed that, we can do the problem itself. The
headings will look like this:

P	Q	R	~R	Q ∧~R	~(Q ∧~R)	[P ∧ ~(Q ∧ ~R)]	~ [P ∧ ~(Q ∧ ~R)]

The next step is to complete the body of the table. Remember the patterns we learned
earlier? The "P" and "Q" and "R" columns are to be filled in using the patterns we
discussed in the previous chapters.

P	Q	R	~R	Q ∧~R	~(Q ∧~R)	[P ∧ ~(Q ∧ ~R)]	~ [P ∧ ~(Q ∧ ~R)]
T	T	T					
T	T	F					
T	F	T					
T	F	F					
F	T	T					
F	T	F					
F	F	T					
F	F	F					

Now we can complete the rest of the table. Under the ~R column we fill in the
OPPOSITE of each entry of the R column. Thus, when R is true, ~R is false and
vise-versa. To do the next column, Q ∧ ~R, we look at the two headings that are
connected, and use the rule for "and", whose symbol is ∧. The next column is the
negation of the previous column.

Thus, whenever Q ∧ ~R is true, then ~(Q ∧ ~R) is false, and vice versa. We then take the first column (P) and the 6th column, ~(Q ∧ ~R) and use the connective "and". The last column is the negation of the preceding column, and the final table looks like this:

P	Q	R	~R	Q ∧~R	~(Q ∧~R)	[P ∧ ~(Q ∧ ~R)]	~ [P ∧ ~(Q ∧ ~R)]
T	T	T	F	F	T	T	F
T	T	F	T	T	F	F	T
T	F	T	F	F	T	T	F
T	F	F	T	F	T	T	F
F	T	T	F	F	T	F	T
F	T	F	T	T	F	F	T
F	F	T	F	F	T	F	T
F	F	F	T	F	T	F	T

Problem Set #4

For problems 1 – 3, write the negation of each sentence.

1) Ohio is a state. _____

2) A dime is not worth 10 cents. _____

3) There are 60 seconds in an hour. _____

4) and 5) and 6) Complete the following truth tables.

P	Q	~P	Q ∧ ~P	Q ∨ (Q ∧ ~P)

P	Q	R	~R	~Q	R ∧ ~Q	P ∨ ~R	(R∧ ~Q) ∧ (P∨ ~R)

P	Q	R	~Q	~R	~Q ∧ ~R	P ∧ (~Q ∧ ~R)	~ [P ∧ (~Q ∧ ~R)]

For problems 7 – 10, the symbols represent these statements. For each problem, write the sentence in words to show what the symbols represent, and state the truth-value of the statement.

P: North Carolina is a state.

Q: Raleigh is a city in North Carolina.

R: A frog is not an animal.

7) ~P: _____ _____

8) ~Q: _____ _____

9) ~R: _____ _____

10) ~ ~ Q : _____ _____

For problems 11 – 13, write the negation of each statement and the truth-value of that negation.

11) $10 + 10 \div 10 = 2$

_____ _____

12)   Seven is a one-digit prime number.

_____ _____

13)   Most taxicabs are yellow.

_____ _____

CHAPTER 5

Conditionals

The next compound statement we will study is a little different from the conjunctions (**and** statements) and the disjunctions (**or** statements) in that it contains two words, not just one word, in the sentence. This type of compound statement is called an "implication" or a "conditional", and is written in the form "if... then..."

Conditionals, or "if-then" statements, are one of the most often used and studied topics in symbolic logic because they often deal with significant cause and effect relationships. Here are two examples of conditionals.

    1) If it is raining, then there are clouds in the sky.

    2) If a two-digit number is prime, then it is odd.

Notice that in each case, the "connective words" are the "if" and the "then". The other parts of the conditional are called the hypothesis and the conclusion. The words or symbols that follow the "if" part of the conditional are called the hypothesis, and the words or symbols that follow the "then" part of the conditional are called the conclusion. Thus, in the first of the two conditionals listed above, the words "it is raining" would be the hypothesis, and the words "there are clouds in the sky" would be called the conclusion. As we did with conjunctions and disjunctions, we assign letters to the parts of a conditional and a symbol as the connective. Do you recall that $P \wedge Q$ is read "P and Q", and that $P \vee Q$ is read "P or Q"? The symbol for an implication or conditional is $\rightarrow$ and, if we assign P as standing for "It is raining" and Q means "there are clouds in the sky" then the implication, "If it is raining, then there are clouds in the sky," would be written in symbolic form as: $P \rightarrow Q$. Some conditionals or "if-then" statements can be written in a different form, requiring us to revise it slightly to be able to see the hypothesis and the conclusion. For example, each of the following are conditionals, but each is written in a slightly different form.

In each example, I have given a statement that is a conditional, but not written in the typical "if-then" form, then I have rewritten it in the standard form.

Example # 1

1) You can be on the team, provided you keep up with your homework.

2) If you keep up with your homework, then you can be on the team.

Example # 2

1) Spencer gets a headache whenever he forgets his glasses.

2) If Spencer forgets his glasses, then he gets a headache.

Although the word order may vary, the hypothesis is always written first when using symbols. For instance, we could have P stand for, "The water temperature is above 80 degrees" and Q stand for, "I will go swimming." Two different ways of writing P → Q would be:

1) If the water temperature is above 80 degrees, then I will go swimming.

2) I will go swimming if the water temperature is above 80 degrees.

Both of these sentences say the same thing, and in both cases the hypothesis is, "The water is above 80 degrees," even though the word order is not the same.

Establishing a truth table for conditionals is a little more complicated because not all of us have as common of an understanding of this form of compound statement as we did with conjunctions and disjunctions.

We will now establish our first truth table for a conditional. Let P serve as the symbol for the hypothesis and Q serve as the symbol for the conclusion. Thus, if your teacher told you: "If you get an A on your final exam, then you will get an A for the year," P is: "You get an A on the final exam", and Q is: "You get an A for the year." Now, let us consider all possible truth-values for the hypothesis and the conclusion. We will do that in a truth table, listing all possible values for the hypothesis and the conclusion using the same order established for the other connectives. Remembering that the top row of every truth table is reserved for the headings, we will begin the establishment of this truth table in the following manner:

P	Q	P → Q
T	T	
T	F	
F	T	
F	F	

Now, in the first row of the body of the truth table, the hypothesis, "You get an A on the exam," is True, and the conclusion, "You get an A for the year," is also true, so your teacher's comment that, "If you get an A on your final exam, then you will get an A for the year," was accurate and honest, so the first row under P → Q is True.

P	Q	P → Q
T	T	**T**
T	F	
F	T	
F	F	

Now, let us consider the second row of the truth table. The statement P is true, and the statement Q is false in that row. So, it is true that you do get an A on the final exam, but you do not get an A for the course (Q is false), thus, your teacher did not tell you the truth, and the conditional is false!

P	Q	P → Q
T	T	T
T	F	**F**
F	T	
F	F	

Row three of the body of this truth table is the most difficult to explain and usually the most difficult to understand. In row three, the hypothesis is False, meaning that you did not get an "A" on the final exam, but the conclusion is true, meaning that you did get an "A" for the year. So now the question arises, was your teacher being truthful and honest when he/she said, "If you get an A on your final exam, then you will get an A for the year"? Well, technically, yes, your teacher was being honest. He/she merely said that you would get an A for the year if you got an A on the exam. He/she did not say the ONLY way for you to get an A for the year was to get an A on the exam... The teacher did not say what would happen if you got a "B" on the exam! So it is possible that you could get a final average of an "A" without getting an A on the exam. It is also possible that you might not, which would be row four of the table.

P	Q	P → Q
T	T	T
T	F	F
F	T	**T**
F	F	

So, as implied in the last sentence of the preceding paragraph, the fourth and final row of the truth table would have to be true also. That means, since in the fourth row the hypothesis is False (You did not get an A on the final exam) and the conclusion is false (You did not get an A for the year), and thus, your teacher was being honest and truthful when she/he said, "If you get an A on your final exam, then you will get an for the year." You did not get an A on the exam, and you did not get an A for the year! We now have the completed truth table and it looks like this:

P	Q	P → Q
T	T	T
T	F	F
F	T	T
F	F	T

As a summary statement, we could say that the **only** time a conditional is false, is when the hypothesis is true and the conclusion is False.

As I stated earlier, conditionals are one of the most important concepts in this logic unit. There are several reasons why they are so important. First, it will just amaze you (I hope) as you become alert to seeing "if...then..." statements in the news, on TV, in books, in daily conversation, in advertising, in the newspapers and in teachers' lectures or in comments that teachers write! *If-then* statements seem to just pop up everywhere! Look for them! The second reason that *if-then* statements are so important, is that they are often misused and can be written in such a way as to imply something to be true that is not true. Discussions that involve arguments or attempts to persuade can be written using *if-then* statements in very misleading ways! We have already learned about *negations*. When a person reverses the hypothesis with the conclusion or negates one or the other or both, a new *if-then* statement is formed. This new statement may or may not have the same truth-values as the original conditional. When an attempt is made to substitute a similar (but different) conditional for the original one, there exists a very real possibility for faulty logic! Be alert and be aware! There are many statements that are implications that might seem to be able to be substituted for other implications without loss of generality or meaning, but by creating truth tables, we can be assured that two reasonably similar implications are or are not equivalent. This will be the focus of the next chapter. Equivalent expressions must have truth-values that are the same in every row! Note the importance of this statement when you attempt problems number 4 and number 5 in the following problem set.

Problem Set # 5

1) If you were to create a truth table for ~ (P ∧ ~Q) ∨ (~P ∧ R), then how many headings would there be? _____

2) What would those headings be for ~ (P ∧ ~Q) ∨ (~P ∧ R) ?

_____ _____ _____ _____ _____ _____ _____ _____ _____

3) Create a truth table for ~ (P ∧ ~Q) ∨ (~P ∧ R).

4) Show, by creating two truth tables, that ~P ∧ Q *is* or *is not* logically equivalent to ~ (P ∨ ~ Q).

P	Q	~P	~P ∧ Q
T	T		
T	F		
F	T		
F	F		

P	Q	~Q	P V ~Q	~ (P V ~ Q)

5) In problem number 4, we could have simplified the process by creating **one** truth table, with the last two columns being ~P ∧ Q and ~ (P V ~ Q) and then comparing those two columns. Let's do that now. If all went well, then the last column of each of the above tables should have the same truth-values!

P	Q	~P	~Q	P V ~Q	~P ∧ Q	~ (P V ~ Q)

6) Look at problem number 5 that you just did and consider the very last sentence in the problem explanation.... "If all went well, then the last two columns should have the same truth-values!" That too is a conditional. If we let K stand for "All went well.", and M stand for "The last two columns had the same truth-values.", then make a truth table for K → M.

K	M	

Now, let us analyze each of the four rows of this truth table. I will do row one, so that you will have a model of what it is you are to do. You do a similar explanation for rows 2 and 3 and 4.

Row 1 — In row 1, K and M are both true, so the implication must also be true, because "All went well" (T) and indeed "the last two columns did have the same truth values" (T)...

Row 2 —

Row 3 —

Row 4 —

7) In the introduction to this project, there was the following sentence: "If any of these chapters, or parts of these chapters, does not make sense, then please see your teacher for further explanation." Let's examine that statement (implication) more carefully by writing it symbolically and creating a truth table for it. First, let's assign letters for the statements.

P = Any of these chapters does not make sense.
Q = Parts of these chapters do not make sense.
R = Please see your teacher for further explanation.

We can now write, "If any of these chapters, or parts of these chapters, does not make sense, then please see your teacher for further explanation" symbolically as $(P \lor Q) \rightarrow R$, and create a truth table like this...

P	Q	R	P ∨ Q	$(P \lor Q) \rightarrow R$

Now, making "sense" of each row can be a challenge! For instance, in row one, all three statements are true and the statement (P ∨ Q) → R was also true. So, it was true that some of these chapters did not make sense (P), and it was true that parts of these chapters did not make sense (Q), and it was true that you saw your teacher for extra help (R). I personally think the next to last row of the truth table is the most confusing one, and that you should be sure to study it carefully! It is important to ask the question: "Which rows in the final column are false, and why are they false?" If you are not sure about the answer to that question, then you should talk to someone who is sure to give you some help. (Another implication!)

8)   Draw a truth table for the following statement: "If a number is neither even nor odd, then it is not prime."

    E = A number is even.
    O = A number is odd.
    P = A number is prime.

E	O	P						

For problems 9 – 12, you are given that P and Q are both true and that R is false. Mark each of the following as True or False or can't be determined (CBD).

9)   _____ (P ∧ R) → ~Q

10)   _____ (P ∨ ~R) → Q

11) _____ P → (R ∨ ~Q)

12) _____ P → (~R ∨ Q)

13)   Consider the following implication: "If you are in the Middle
School and currently taking Algebra 1, then you must be in the 7th or 8th grade."
Define variables for each of the four statements, write the compound statement in
symbolic form, and create a truth table for the implication. Note that I have chosen
variables this time that are "connected" to the sentence they represent. Assume that
the original implication is true.

Let MS = _____

Let A = _____

Let S = _____

Let E = _____

MS	A	S	E			

CHAPTER 6

Converse, Inverse, and Contrapositive

Chapter 5 introduced the notion of implications, or conditionals. This chapter looks at variations of implications. There are three common variations of a conditional that can use the same hypothesis and conclusion, but then switch them around in ways that require us to think about what they mean and how to use them. We will consider them one at a time. First, if we take the implication "If a two-digit number is prime, then that number is odd" and assign variables to the hypothesis and conclusion, we get:

P = A two-digit number is prime.
Q = That number is odd.

and the implication would be written P → Q.

If we create a new *If ... then* statement using the same simple sentences for P and Q, but <u>reversing</u> the hypothesis and conclusion, we get:

"If a two-digit number is odd, then it is prime"

and we write this as Q → P.

When the hypothesis and conclusion of a conditional are reversed, the new conditional is called the **converse** of the original implication. Note that in this case, the original implication was true, but the converse of the implication is false. That is, it is true that every two-digit prime number is odd, but it is not true that every two-digit odd number is prime! This observation is important. Sometimes a conditional and its converse have the same truth-value, sometimes they do not. In the example just used, they do not, but in the following example they do. At Edgewood Middle School, the grading scale is 90 – 100 is an A. Therefore, the conditional: "If your score is 90% or greater on the test, then you will receive an A on that test," and its converse: "If you received an A on the test, then you scored 90% or greater," have the same truth-value. We must be very careful to not randomly substitute the converse of a statement for the statement itself when debating or trying to "prove" something to be true. Sometimes they can be substituted for one another, and sometimes they cannot! A statement and its converse are not equivalent.

The second of the three "related" implications to a conditional is called the **inverse**. The **inverse** of a conditional is formed by negating the hypothesis and conclusion and forming a new conditional. Thus, the inverse of P → Q would be ~P → ~Q. Using the same two conditionals that we used when explaining converse, let us now look at the inverse of, "If a two digit number is prime, then it is odd." The inverse of that statement would be: "If a two digit number is not prime, then it is not odd." Again, the inverse has a different truth-value than the original conditional. The inverse of any conditional has the same order for the hypothesis and conclusion, but each is negated. Think about what the inverse of, "If your score is 90% or greater on the test, then you will receive an A on that test," would be. Remember, to find the inverse, we just write the negation of the hypothesis and conclusion and form a new implication. The inverse would be: "If your score is not 90% or greater on the test, then you will not receive an A on that test." Here is an example of a statement and its inverse that have the same truth-value. Since we showed in the previous example that the statement and its inverse did not have the same truth-values, we can conclude that it is not necessarily true that a statement and its inverse are equivalent. Two statements are said to be equivalent only if they have the identical truth table results.

The fourth and final variation of a conditional is called the **contrapositive**. The **contrapositive** of a statement takes the hypothesis and conclusion and reverses them and negates them. Thus, the contrapositive of P → Q would be ~Q → ~P, and would be read as "If not Q, then not P." Lets look at the same two examples that we used in explaining both the converse and inverse. Think about how to write the contrapositive of, "If a two digit number is prime, then it is odd." Remember we must both reverse the two parts of the statement and negate them. The contrapositive would be: "If a two digit number is not odd, then it is not prime." In like manner, the contrapositive of, "If your score is 90% or greater on the test, then you will receive an A on that test," is, "If you did not receive an A on the test, then your score was not 90% or greater." Notice that in both cases, the implication and its contrapositive had the same truth-values! In order to establish the equivalence or lack of equivalence of a statement with its converse and inverse and contrapositive, we need to construct a truth table and analyze it. Consider the following truth table.

P	Q	~P	~Q	$P \rightarrow Q$ (Implication)	$Q \rightarrow P$ (Converse)	$\sim P \rightarrow \sim Q$ (Inverse)	$\sim Q \rightarrow \sim P$ (Contrapositive)
T	T	F	F	T	T	T	T
T	F	F	T	F	T	T	F
F	T	T	F	T	F	F	T
F	F	T	T	T	T	T	T

Notice from the table that the **implication** and the **contrapositive** have truth-values that are exactly the same in every row. In like manner, the **converse** and the **inverse** also have truth-values that are identical in every row. We say that a statement and its **contrapositive** are logically equivalent (or interchangeable) and we can also say that a **converse** and **inverse** are logically equivalent (or interchangeable). Note that while we can talk about a statement as something that exists in its own right, it is not possible to discuss the inverse or converse or contrapositive except in relationship to a conditional. That is, to talk about an inverse... we must ask the question relative to some given conditional.

Problem Set # 6

1) Give the converse, inverse, and contrapositive of:

if $X = 7$ then $X^2 = 49$.

Then state the truth value of each.

Converse: _____ _____

Inverse: _____ _____

Contrapositive: _____ _____

For problems 2 through 10, write your answer in symbolic form.

2) What is the converse of ~B → ~V ? _____

3) What is the converse of J → ~H ? _____

4) What is the inverse of W → T ? _____

5) What is the inverse of ~G → M ? _____

6) What is the contrapositive of ~ V → K ? _____

7) What is the contrapositive of ~ T → ~X ? _____

8) What is the converse of the inverse of W → V ? _____

9) What is the inverse of the converse of ~R → T ? _____

10) What is the contrapositive of the contrapositive of P → Q ? _____

For problems 11 – 14, write your response in words and state if the given response is true or false.

11) Consider the statement: "If an integer ends in zero, then it is divisible by ten."

A) Converse: _____ _____

B) Inverse: _____ _____

C) Contrapositive: _____ _____

12) Consider the statement: "If a figure has four sides, then it is not a triangle."

A) Converse: _____ _____

B) Inverse: _____ _____

C) Contrapositive: _____ _____

13) Consider the statement: "If a number is odd, then it is either positive or negative."

A) Converse: _____ _____

B) Inverse: _____ _____

C) Contrapositive: _____ _____

14) Consider the statement: "If it isn't broken, then it doesn't need to be fixed."

A) Converse: _____ _____

B) Inverse: _____ _____

C) Contrapositive: _____ _____

15) Notice that the statement in problem number 13 was a compound implication. That is, the hypothesis was a simple statement, but the conclusion contained the connective "or". Thus, if we were to write this statement symbolically, it could look like this:

Let O = The number is odd.
Let P = The number is positive.
Let N = The number is negative.

And the statement would be written as: O → (P ∨ N). Write the converse, inverse, and contrapositive in words, and use symbols to write the converse, inverse and contrapositive and construct a truth table for all four statements.

Statement = If a number is odd, then it is either positive or negative.

Converse = _____

Inverse = _____

Contrapositive = _____

O	P	N	~O	~P	~N	P∨N	~(P∨N)	The statement itself O → (P ∨ N)	The converse _____	The inverse _____	The contrapositive _____

Each of problems 16 – 18 gives either the converse or inverse or contrapositive of a conditional. State what the conditional must be, and the other two statements that are missing.

16) ~ M → K is the converse of the conditional.

Conditional = _____

Inverse = _____

Contrapositive = _____

17) D → ~ R is the contrapositive of the conditional.

Conditional = _____

Converse = _____

Inverse = _____

18) ~W → ~M is the inverse of the conditional.

Conditional = _____

Converse = _____

Contrapositive = _____

For each problem in 19, 20, and 21, construct truth tables for each expression to determine if the two given statements are or are not equivalent.

19) W ∧ ~ U : ~U ∧ ~ ~ W   *__Are__* or *__are not__* equivalent (Circle one)

W	U					

20) (P ∨ Q) → ~ R : R → ~ (Q ∨ P)
     __Are__ or *__are not__* equivalent (Circle one)

P	Q	R					

21) ~ (Q ∧ ~W) : W ∨ ~ Q   *Are* or *are not* equivalent (Circle one)

W	Q					

22) Your algebra teacher said to you, "If you don't do all the extra credit problems, then you will not get an A for the quarter." So, you indeed do all the extra credit, but yet your algebra teacher does not give you an A for the quarter. When this happens, you get angry and claim that your teacher had lied to you. Your teacher said he did not lie! Who was right? _____

Support your answer with a Truth Table. Be sure to identify the variables.

23) Your coach said, "If you don't lift weights, you'll be sorry." So, not wanting to be sorry, you lifted weights. But, you pulled a muscle while lifting weights and you were sorry you did it. You went back to your coach and complained that he was wrong about what he said. Construct a truth table (being sure to define the variables appropriately) for the implication and explain why the coach was or was not wrong with what he had said.

```
_____
_____
_____
_____
_____
```

The coach **_was_** or **_was not_** wrong because..... _____
      (circle one)

_____

_____

24) Let P = Paul got an A in algebra.
    Let M = Madison got an A in algebra.

Give the meaning of each of the following five statements.

   a) ~P $\lor$ ~M _____

   b) ~ (P $\lor$ M) _____

   c) ~P $\land$ ~M _____

   d) ~ (P $\land$ M) _____

   e) ~ ~ ~ P _____

25) "If two lines intersect, then they are not parallel," is the contrapositive
    of what statement? _____

_____

26) "If two lines do not intersect, then they are parallel," is the converse of
   what statement? _____

   _____

27) "If two lines are not parallel, then they do intersect," is the inverse of
   what statement? _____

   _____

28) Which (if any) of these four statements are true?

   a) _____ If two lines are parallel, then they do not intersect.

   b) _____ If two lines do not intersect, then they are parallel.

   c) _____ If two lines are not parallel, then they do intersect.

   d) _____ If two lines intersect, then they are not parallel.

CHAPTER 7

Biconditionals

As we continue to add more symbols, it is important to remember that the new symbols are indeed just another "operation" in symbolic logic, and that the old symbols that we have already learned remain the same. A quick review of the symbols we have learned thus far:

1) $\vee$ means "or" and is the connective called a disjunction

2) $\wedge$ means "and" and is the connective called a conjunction.

3) $\sim$ means "not" and is called a negation.

4) $\rightarrow$ is the symbol that is used for a conditional or implication and is read "if... then."

The fifth, and final, symbol that we will learn in this project is used for what is called a "biconditional." A biconditional is written $P \leftrightarrow Q$ and can be read several ways:

P if and only if Q

or

If P then Q, and if Q then P

or

P implies Q, and Q implies P.

Note that there are two conditions placed on the variables. Because it is an "and" statement, both $P \rightarrow Q$ **and** $Q \rightarrow P$ must be true for the biconditional to be true. Here is a truth table summarizing that information:

P	Q	P → Q	Q → P	(P → Q) ∧ (Q → P)
T	T	T	T	T
T	F	F	T	F
F	T	T	F	F
F	F	T	T	T

Remembering that mathematicians and logicians are lazy (or was it efficient?), rather than writing (P → Q) ∧ (Q → P) every time we wished to write a biconditional, we shorten the reading and the writing of the biconditional in the following manner. We shorten the reading by using the words "if and only if," and we shorten the writing by introducing the symbol ↔ , which will replace the compound conjunction of the two implications. Thus, P ↔ Q and (P → Q) ∧ (Q → P) are two ways of writing the same thing! You will note in the truth table, that the biconditional is true only when P and Q had the same truth-values. That is, for the biconditional to be true, the two connected statements must be both true or both false. When one of the two statements connected with the ↔ symbol is true and the other is false, then the biconditional is false. From this information, we could conclude that a truth table for a biconditional would look like this:

P	Q	P ↔ Q
T	T	T
T	F	F
F	T	F
F	F	T

In that never-ending attempt at efficiency, mathematicians have made an abbreviation for "if and only if." That abbreviation is: "iff". Therefore, if P represents "a number is divisible by 10," and Q stands for "a number ends in zero," we could shorten the biconditional: "If a number is divisible by 10, then it ends in zero and if a number ends in zero, then it is divisible by 10," to "A number is divisible by 10 iff it ends in zero." One might wonder if the teacher who said, "If you get an A on the exam, then you will get an A for the year," might really have been thinking and/or wanting to convey, "You will get an A for the year if and only if you get an A on the exam." The use of a conditional instead of a biconditional is a common error in reasoning.

Problem Set # 7

Mark each of problems 1 – 10 as true or false.

1) _____ If a number ends in 6, then it is even.

2) _____ If a number is even, then it ends in six.

3) _____ A number is even iff it ends in 6.

4) _____ If a figure is a square, then it has four sides.

5) _____ If a figure has four sides, then it is a square.

6) _____ If $x + 1 = 7$, then $x = 6$.

7) _____ If $x = 6$, then $x + 1 = 7$.

8) _____ $x + 1 = 7$ iff $x = 6$.

9) _____ If $x = 3$, then $x^2 = 9$.

10) _____ If $x^2 = 9$, then $x = 3$.

Consider the following given information for problems 11 – 18:

   M = I will work on Monday.
   T = I will work on Tuesday.

11) Write the following statement with symbols: "If I do not work on Monday, then I will work on Tuesday." _____

12) Write the following symbolic statement in words.

$[(M \vee T) \wedge \sim T] \rightarrow M$ _____

13) Draw a truth table for   [(M ∨ T) ∧ ~ T] → M

M	T				

14) From your response in problem # 12, and from the truth table in problem # 13, what, if anything, can we conclude about M? _____

15) Write ~ (M ∧ T) in words _____

16) Write ~M ∨ ~T in words _____

17) Draw a truth table for ~ (M ∧ T)

M	T		

18) Draw a truth table for ~M ∨ ~T

M	T			

19) Compare your truth tables in problems 17 and 18. State an observation and the significance of that observation. _____

_____

20) In chapter 3 of this project, there was the sentence: "A conjunction is true if and only if both of the connected statements are true." How is that sentence similar and different from, "A conjunction is true if both of the connected statements are true"?

_____

_____

Addendum to Problem Set # 7

For each of the problems 1–27, determine if the expression is or is not a "statement". If it is a statement, state if it is true or false. If it is not a statement, state why. Some reasons could be that it is an open sentence, that it is an opinion, that it is a command, that it is a question, or that it is a phrase.

You might recall that the above sentences were the directions for the first set of problems in Chapter 1 of this project…. Now that we have studied and learned about conjunctions and disjunctions and negations and implications and their converse, inverse, and contrapositive, let's use those same set of directions to determine if the following expressions are or are not statements. Remember, if you determine that it is a statement, state if it is true or false. If it is not a statement, state why.

1) The square of an even integer is an even integer, or the square of an odd integer is an odd integer. _____

    _____

2) All two-digit prime numbers are odd, or all two-digit odd numbers are prime. _____

3) If a two-digit integer is prime, then it is odd. _____
    _____

4) If a two-digit integer is odd, then it is prime. _____
    _____

5) If you do not finish this project, then you will die! _____
    _____

Problems 6 – 9 refer to the following sentence: P = If a number is odd, then it is positive or negative.

6) The converse of P. _____

7) The inverse of P. _____

8) The contrapositive of P. _____

9) The statement, P. _____

10) It is false that the statement Q and the statement ~Q never have the same
truth-value. _____

11) Every statement is true or false. _____

12) Every statement that is not true, is false. _____

13) If X and Y are two prime numbers, and the sum of X and Y is prime, then at
least one of the two numbers must be even. _____

_____

14) No even number is divisible by an odd number. _____

_____

15) No odd number is divisible by an even number. _____

_____

16) If a number is divisible by 6, then it is even. _____

_____

17) The converse of the previous problem. _____

_____

18) The inverse of problem number 16. _____

_____

19) The contrapositive of problem number 16. _____

_____

20) Can you name a two-digit prime number that is odd? _____

_____

21) Can you name a two-digit odd number that is not prime? _____

_____

22) If you think these problems are easy, then you are smart! _____

_____

23) If it is false that ~Q is true, then it is true that ~ (~Q) is false. _____

_____

24) The day after Monday is either Tuesday or Wednesday. _____

_____

25) The day after Wednesday is either Friday or Saturday. _____

_____

26) If today is Wednesday, then yesterday was Tuesday. _____

_____

27) All books in the public library are worth reading. _____

_____

CHAPTER 8

Tautologies

Tautologies are the logical equivalent to an equation that is always true. For instance, suppose I give you the following equation to solve:

$$2(3x + 5) + 2 = 3(2x + 4).$$

When we solve that equation by using the distributive property, combining terms, getting the variables on one side, and solving for x, we find that the statement is always true, regardless of the value of x. This means the solution is the universal set of numbers. In logic, instead of giving you an equation to solve (as in algebra), I would give you a problem and ask you to create a truth table. If the truth table turns out to have all "T's" in the final column, then the expression we were "solving" was a tautology. Consider the following example: Draw a truth table for:

$$\sim (M \wedge T) \leftrightarrow (\sim M \vee \sim T).$$

You might note that the left side of this equivalence was problem number 17 in problem set number 7, and the right side of the equivalence was problem number 18. Let's draw the truth table.

M	T	~M	~T	M ∧ T	~ (M ∧ T)	~ M ∨ ~T	~ (M ∧ T) ↔ (~ M ∨ ~T)
T	T	F	F	T	F	F	T
T	F	F	T	F	T	T	T
F	T	T	F	F	T	T	T
F	F	T	T	F	T	T	T

Notice that the last column (the problem itself) is all True. This means regardless of the conditions, $\sim (M \wedge T) \leftrightarrow (\sim M \vee \sim T)$ is always True. To demonstrate that an expression is or is not a tautology, you can create a truth table and check to see if the statement is or is not always true. If it is, then it is a tautology. Tautologies are often used to develop strong arguments. It is important to note the difference between a tautology and an equivalence. Two statements are said to be equivalent if they say the same thing in two different ways. For instance, ~P and ~ ~ ~ P are equivalent. Also, ~ (M ∧ T) and ~ M ∨ ~T are equivalent. There are, in fact, an infinite number of logically equivalent statements, but just a few of them important in the reasoning/ thinking process.

One way to test to see if two statements are equivalent is to compare the two statements using the biconditional. If the two connected statements have the same truth values, then the biconditional will be true in every case! So, the fact that an expression is or is not a tautology can always be verified by setting up a truth table.

Problem Set # 8

1) Show that ~Q → P **_is_** or **_is not_** logically equivalent to P ∨ Q. If they are logically equivalent, state a tautology to show this equivalence.

P	Q	

They **_are_** or **_are not_** logically equivalent? _____

For problems 2 – 4, formulate a logically equivalent statement, using only P, Q, ~, ∧, ∨, → and ↔. You may use as many parentheses as you need. Use a truth table to explain why the statements are logically equivalent.

2) (P ∧ ~Q) ∨ ~P   _____

P	Q	

3) P ↔ ~ Q   _____

P	Q	

4) ~ P ∧ ~ Q _____

P	Q	

5) Explain why P → Q ∧ R is an ambiguous statement. (Hint: It has something to do with binary operations.)

6) Draw a truth table for P → [~P → (P → { ~P → Q})] and explain why it is or is not a tautology.

P	Q	

Cumulative Review

1) Name four ways you could logically and accurately respond to the question, "Do you want soup or salad with your dinner?"

    1)

    2)

    3)

    4)

2) Draw a truth table for $(P \lor Q) \rightarrow R$

P	Q	R	

3) Draw a truth table for $(P \rightarrow R) \wedge (Q \rightarrow R)$

P	Q	R	

4) If $\sim P \wedge Q$ is True, then P is _____.

    A) True   B) False   C) Cannot be determined

5) If $\sim R \vee M$ is True, and M is False, then R is _____.

    A) True   B) False   C) Cannot be determined

6) If an implication is False, then its hypothesis is _____.

    A) True   B) False   C) Cannot be determined

7) If the contrapositive of a statement is False, then the inverse of the statement is _____.

    A) True   B) False   C) Cannot be determined

8) If P and Q are True, then $\sim Q \rightarrow P$ is _____.

    A) True   B) False   C) Cannot be determined

9) If P and Q are True, then ~Q → ~P is _____.

    A) True   B) False   C) Cannot be determined

10) If P and Q are True, then P → ~Q is ____.

    A) True   B) False   C) Cannot be determined

11) Your algebra teacher said, "You will get an A in algebra if you have an above 90% average on your quizzes, or you do all your homework and you do not miss more than two classes." Which of the following students will definitely get an A in algebra?

_____ Stephen got a 97% average on his quizzes, did all his homework, and missed one class.

_____ Vincent got a 95% average on his quizzes, did all his homework, and missed seven classes.

_____ Ellen got a 92% average on her quizzes, did not do all her homework, and missed one class.

_____ Carolyn got a 99% average on her quizzes, did not do all her homework, and missed three classes.

_____ David got a 89% average on his quizzes, did all his homework, and missed no classes.

_____ Sammy got a 77% average on his quizzes, did all his homework, and missed 14 classes.

_____ Kyle got a 44% average on his quizzes, did not do all his homework, and missed two classes.

_____ Elizabeth got an 89% average on her quizzes, missed one homework, and missed three classes.

12) The next year, the teacher gave the same directions, but changed the position of one comma, giving the following statement: "You will get an A in algebra if you have an above 90% average on your quizzes or you do all your homework, and you do not miss more than two classes." Which of the following students will definitely get an A in algebra?

_____ Stephen got a 97% average on his quizzes, did all his homework, and missed one class.

_____ Vincent got a 95% average on his quizzes, did all his homework, and missed seven classes.

_____ Ellen got a 92% average on her quizzes, did not do all her homework, and missed one class.

_____ Carolyn got a 99% average on her quizzes, did not do all her homework, and missed three classes.

_____ David got a 89% average on his quizzes, did all his homework, and missed no classes.

_____ Sammy got a 77% average on his quizzes, did all his homework, and missed 14 classes.

_____ Kyle got a 44% average on his quizzes, did not do all his homework, and missed two classes.

_____ Elizabeth got an 89% average on her quizzes, missed one homework, and missed three classes.

13) Give the converse, inverse, and contrapositive of, "If I don't water the plants, then they will die," ***and state the truth-value of each***. Assume the original implication is true.

Converse = _____

Inverse = _____

Contrapositive = _____

For problems 14 – 16, assume that statements one and two are both true, and write a conclusion, if possible. If it is not possible to form a conclusion, write CBD (Can't Be Done).

14) If I am late for school, then my teacher will be mad.
     I am not late for school.

_____

15) I like algebra or French.
     I do not like French.

_____

16) I like algebra or French.
     I like algebra.

_____

17) Draw a truth table for: A $\leftrightarrow$ {A $\leftrightarrow$ [Q $\lor$ (H $\land$ M)]}

Q	H	M	A	

18) Draw a truth table for: A ↔{[(Q∨H) ∧ M] ↔A}

Q	H	M	A	

19) Please give some feedback to your teacher as to what you thought of this project, and how it can be improved when used the next time. _____

_____

_____

_____

_____

CONCLUSION

My hope is now that you have finished, or nearly finished this project, you have a better understanding of what it means to "think logically," and a clearer notion about how to do that. I believe there is an excellent chance that you will be more aware of how you use language, how language can be (and is) misused, and that you have strengthened your ability to make your ideas clearer to yourself and others. In virtually every setting where two or more people are communicating, there is a need to assess what assumptions are made. Education is, I believe, a never-ending search for truth, and it seems clear to me that an understanding of the concepts presented in this project will enhance that search. As we increase our awareness of the power of deductive reasoning, we will be on the road to minimizing the errors in reasoning we see every day, in TV advertisements, in teacher's lectures, in books, in newspapers, in speeches, and in normal everyday conversations. If… then… statements are embedded in virtually every form of communication. Mastery of understanding conditionals, their converses, inverses, and contrapositives, is crucial to the thought process. The clearness of thought associated with logical thinking will help us develop the ability to detect illogical inferences and to not be inappropriately swayed to believe something based on logic that we recognize as faulty. The bottom line is that a thorough understanding of the laws of logic will help to foster our ability to communicate effectively, not only in mathematics, but also in virtually every endeavor. Precision of thought and clear reasoning are, I believe, lofty but attainable goals that we should all aspire to. The concluding of this project does not indicate the end of your study of logic, just to the end of this step in the process. If we have learned anything from this project, (Do I see an implication in the making?) then we should realize that logical thinking is a learned skill which must, like all learned skills (for example: playing the piano, hitting a golf ball, and taking good notes) be used and refined and nurtured to grow and become stronger and more useful to us. I hope that your thoughtful completion of this project has been an important step in this never-ending process. The next section of this text is called the *Binomial Expansion*. There are some powerful connections between this logic project and the binomial expansion project.

BINOMIAL EXPANSION PROJECT

Introduction

Is there *really* some use for $(A + B)^2$?

Is there an easier way to find $(A + B)^8$?

What do these two questions have in common?

What happens if the terms of the binomial are binomials themselves?

Is there such a thing as a "trinomial expansion"?

The answers to these and other questions will, I hope, become evident as you complete this project. There are a variety of methods for teaching/learning about the binomial expansion. As was true for part one of this text, your teacher has several options and will decide how this project will be presented. While this project is shorter than the Symbolic Logic project, it will probably take about the same amount of time to complete, as many of the problems tend to have longer solutions and answers. When you do the problems on the assignment page, be sure to show all work and express all answers in simplest form. A prerequisite for the study of this project is the ability to multiply binomials. Often this is taught as the "FOIL" method, where FOIL stands for **F**irst, **O**uter, **I**nner, **L**ast. This short cut allows us to be able to multiply two binomials in our heads. Unfortunately, it only works for the product of two binomials. If we want to multiply more than two binomials or if we want to multiply polynomials with more than two terms, the FOIL method does not work. The binomial expansion does! Learning how to expand binomials has implications and applications much beyond the scope of this project. As was true for the previous project, support is available from my web site and via email, as well as from your teacher.

Note the following patterns:

$(A+B)^1 =$ $A + B$

$(A + B)^2 =$ $A^2 + 2AB + B^2$

$(A + B)^3 =$ $A^3 + 3A^2B + 3AB^2 + B^3$

$(A + B)^4 =$ $A^4 + 4A^3B + 6A^2B^2 + 4AB^3 + B^4$

$(A+B)^5 =$ $A^5 + 5A^4B + 10A^3B^2 + 10A^2B^3 + 5AB^4 + B^5$

$(A - B)^6 =$ $A^6 - 6A^5B + 15A^4B^2 - 20A^3B^3 + 15A^2B^4 - 6AB^5 + B^6$

These examples suggest:

1) The number of terms in the expansion on $(A + B)^N$ is $N + 1$.

2) If the binomial is a sum, all the terms in the expansion are added. If the binomial is a difference, the terms are alternately added and subtracted, with the even numbered terms being subtracted.

3) The coefficient of the first term is 1.

4) The exponent of "A" in the first term is "N". The exponent of "A" in any term after the first term is one less than the exponent of "A" in the preceding term.

5) The coefficient of any term (except the first term) is the product of the coefficient of the preceding term and the exponent of "A" in the preceding term, divided by the number of the preceding term.

6) The sum of the exponents of "A" and "B" in each term is "N".

If you understand the information on the previous page, then you should be able to carry (A+B) to any power. For instance, can you find $(A+B)^{14}$? Here is the expansion of $(A+B)^N$.....

$$(A+B)^N = A^N + \frac{N}{1}A^{N-1}B^1 + \frac{N}{1}\frac{N-1}{2}A^{N-2}B^2 + \frac{N}{1}\frac{N-1}{2}\frac{N-2}{3}A^{N-3}B^3 \text{ .... } B^N$$

The 7th term would be:

$$\frac{N}{1}\frac{N-1}{2}\frac{N-2}{3}\frac{N-3}{4}\frac{N-4}{5}\frac{N-5}{6}A^{N-6}B^6$$

or......

$$\frac{(N)(N-1)(N-2)(N-3)(N-4)(N-5)}{6!}A^{N-6}B^6 \text{ ......}$$

The use of "!" is called *Factorial Notation* and indicates the product of successive natural numbers less than or equal to the given number... Thus...

$$6! = (6)(5)(4)(3)(2)(1)$$

and

$$N! = (1)(2)(3)(4) ...N \text{ (where N is any natural number)}$$

By definition, $0! = 1$

From this information, we can conclude that the "Rth" term of $(A+B)^N$ is:

$$\frac{N(N-1)(N-2)(N-3)...[N-(R-2)]}{(R-1)!} \; A^{[N-(R-1)]} \; B^{(R-1)}$$

This is an important formula! — Do you know what "N" and "R" and "A" and "B" stand for?

EXAMPLE: Expand completely $(2x+3)^7$. I will do this in two steps, first of all in the un-simplified form, then in the simplified form.

$$(2x+3)^7 = (2x)^7 + \frac{7}{1}(2x)^6(3)^1 + \frac{(7)(6)}{(2)(1)}(2x)^5(3)^2 + \frac{(7)(6)(5)}{(1)(2)(3)}(2x)^4(3)^3 +$$

$$\frac{(7)(6)(5)(4)}{(1)(2)(3)(4)}(2x)^3(3)^4 + \frac{(7)(6)(5)(4)(3)}{(1)(2)(3(4)(5)}(2x)^2(3)^5 + \frac{(7)(6)(5)(4)(3)(2)}{(1)(2)(3)(4)(5)(6)}(2x)(3)^6+$$

$$\frac{(7)(6)(5)(4)(3)(2)(1)}{(1)(2)(3)(4)(5)(6)(7)}(3)^7$$

Notice that what I've done is to take the formula on the top of page 104 and plug in specific values for "A", "B", and "N". In this case, "A" (which stands for the first term of the binomial) is "2x" and "B" (which stands for the second term of the binomial) is "3" and N (which stands for the exponent that the binomial is being carried to) is "7"... The final simplified form is:

$$(2x+3)^7 = 128x^7 + 1344x^6 + 6048x^5 + 15120x^4 + 22680x^3 + 20412x^2 + 10206x + 2187$$

Another example: Find the 6th term of $(x+2y)^{12}$. Taking the formula at the top of page 105 we get:

$$\text{6th term} = \frac{(12)(11)(10)(9)(8)}{(1)(2)(3)(4)(5)} \, x^7 \, (2y)^5 = 25{,}344 \, x^7 y^5$$

A technique that is often used in advanced mathematics is the substitution of one number for another to make the problem easier to solve. An example would be if we wanted to expand $(A+B+C)$ to the third power. One way to do this problem would be to write $A+B+C$ three times, then multiply the first two trinomials together, and then finally take that answer and multiply it by $A+B+C$ again. A different (and significantly easier) way to do this problem would be to substitute the $A+B$ with an "X", and then we would be cubing a binomial rather than a trinomial. Cubing a binomial can be done in our heads, if we understand the binomial expansion. With this substitution, we now have $(X+C)^3$, which would be:

$$X^3+3X^2C+3XC^2+C^3.$$

Now, we can take that answer and replace the X with the $A+B$ and get that same polynomial, only this time with the original variables. That answer would be:

$$(A+B)^3+3(A+B)^2C +3(A+B)C^2+C^3$$

which can be simplified by use of the binomial expansion on the first and second terms and we would get:

$$A^3+3A^2B+3AB^2+B^3+3A^2C+6ABC+3B^2C+3AC^2+3BC^2+C^3.$$

There are implications of this technique and this polynomial that go far beyond the scope of this project, but can be found in the January 2000 issue of the *Mathematics Teacher*. It would be worth your while to look this up and read about it, especially with the expertise you have developed by doing these two projects!

ASSIGNMENTS

(1) As you work on this project, always keep "Addendum #1" and the Time Sheet available and keep it up to date as you proceed.

(2) Explain <u>in detail</u> each of the following items:

    (A) How many terms there are in the expansion of $(a+b)^n$.
    (B) The pattern in the numerical coefficients in the expansion.
    (C) The difference it makes if the binomial is a sum or difference.

(3) Check out at least 3 websites about Pascal's Triangle. List the two websites you found most interesting and pick one and identify specifically the most complicated information in that site that you feel is still within your level of expertise. Explain what that was and why you chose that particular piece of information.

(4) Check out at least 5 websites about binomials. List the 3 websites you found most interesting and pick one from that group and identify specifically the most complicated information in that site that you feel is still within your level of expertise. Explain what that was and why you chose that particular piece of information.

(5) Complete Addendum # 2 worksheet.

(6) Complete Addendum #3 worksheet.

(7) Complete Addendum #4 worksheet.

ADDENDUM # 1

Binomial Expansion Project Time Sheet

Name = _____

Date	Beginning time. (Rounded to nearest 5 min.)	Ending time. (Rounded to nearest 5 min.)	Number of minutes worked this session.	Total number of minutes worked thus far.	Brief, but **specific**, description of what you did this session.

ADDENDUM # 2

1) In the formula:

$$\frac{N(N-1)(N-2)(N-3)...[N-(R-2)]}{(R-1)!} A^{[N-(R-1)]} B^{(R-1)}$$

State what each variable stands for.

N = _____ R = _____

A = _____ B = _____

2) What is the numerical coefficient of the next to last
   term of $(x+y)^{488}$? _____

3) Give the 8th term of $(3x^{15} - y^8)^9 =$ _____

4) Give the first three terms of $(a^7 + 2b^{10})^{11}$

_____ + _____ + _____

5) Find the 41st term of $(2x^{10} + y^{17})^{45} =$ _____

6) Give the **last** three terms of : $(3x^{11} - y^4z^6)^{11} =$

7) Find the middle term of : $\left(\frac{x}{2} + \frac{2}{x} \right)^8 =$ _____

ADDENDUM #3

1) $(A+B)^{75}$ when expanded has _____ terms.

2) The 51st term of $(7x^{10} + b^4d)^{50}$ is _____ .

3) What is the sum of the exponents of the variables in the 54th term of $(A-B)^{70}$?

_____

4) What is the numerical coefficient in the next to last term of

$(2X - Y)^{4803}$? _____

5) Explain the similarities and differences between carrying $(2x - r)$ to the 30th power and carrying $(2x+r)$ to the 30th power.

6) $\dfrac{(20)(19)(18)(17)...(8)(7)(6)}{(15)(14)(13).....1} \, (2x)^5(y^{10})^{15}$ is the _____ term

of $(2x +y^{10})^{20}$.

7) Give the middle term of $(\dfrac{1}{4}x + 4y^{12})^{16}$ = _____

8) Use the substitution technique discussed on page 103 to simplify $(2M+N+3R^3)^4$.

ADDENDUM # 4

1) If N = 12 and R = 4 , show that the following equation is or is not true. Comment on what your solution might mean.

$$\frac{N!}{(N-R+1)!(R-1)!} = \frac{N(N-1)(N-2)(N-3)...[N-(R-2)]}{(R-1)!}$$

2) Give the last three terms of $(5b^{40} + x^{11})^{30}$ ...

3) Give the first four terms of $(a^{77} - 2rxa^8b^{10})^{27}$ ...

4) Expand $(\frac{1}{3}y^{11} - 6x^4y)^7$ completely.

SUMMARY, CONCLUSIONS, AND CONNECTIONS

In chapter 2 of the "Logic Project", there was reference to what could happen when two coins are tossed, or when three coins are tossed, or more! Now that you have studied the Binomial Expansion, we can approach that set of problems from another perspective. While studying these pages, think back about the same problems when done from the perspective of Truth Tables, and compare the results here with the results in chapter 2 of the previous project. Consider what happens when we carry (H + T) to the second power, using the binomial expansion. We get:

$$H^2 + 2HT + T^2.$$

Notice that the answer is a trinomial, with the exponent of H in the first term being 2, the exponents of both variables in the second term are 1, and the exponent of T in the last term is 2. The fact that it is a trinomial indicates how many different outcomes are possible. Since the coefficient of the middle term is 2, that outcome can occur twice! Also observe that the sum of the coefficients in the three terms is 4. The sum of the coefficients gives us the total number of possible results that could happen if we toss two coins. In other words, there is one way to get two heads, two ways to get one head and one tail, and one way to get two tails.

Think carefully about this question: If you carried (H+T) to the third power (representing three coins), how many total terms would there be in the answer?

$$(H+T)^3 = H^3 + 3H^2T + 3HT^2 + T^3$$

The exponents of the variables tells us how many Heads or Tails there are, and the coefficients of the terms tells us how many ways to get that number of terms. Thus….. In the first term, there is one (coefficient) way of getting three (exponent) Heads (Variable of H) …. In the second term, there are three (coefficient) ways of getting two (exponent) Heads (variable) and one (exponent) Tail (variable). In all, the sum of the coefficients gives us the total number of possible outcomes, and the exponents give us what those outcomes are! This can be summarized in the following table.

Coin #1	Coin #2	Coin #3	
H	H	H	One row with three H
H	H	T	One of the three rows with two H and one T
H	T	H	One of the three rows with two H and one T
H	T	T	One of the three rows with two T and one H
T	H	H	One of the three rows with two H and one T
T	H	T	One of the three rows with two T and one H
T	T	H	One of the three rows with two T and one H
T	T	T	One row with three T

Note, that just like with the Truth Tables in the previous project, there are eight possible combinations with three variables. The same is true with any *binary* distribution. If you do not remember how to count in **base 2**, you should take time out now and look that up on the internet; http://www.itnewb.com/v/Introduction-to-the-Binary-or-Base-2-Number-System is a web site that gives a nice introduction to a binary, or **base 2**, counting system, or ask your teacher. There are many sites about binary number systems, and I would encourage you to look several of them up if you need a refresher. In short, counting in base two uses just two symbols, Zero and One. Just like Truth Tables rely on statements that have just two possible conditions (True or False) and coin flipping has just two possible outcomes (Heads and Tails), there are other binary systems that are important to our world. Computer systems use a binary code, as does electrical wiring. Virtually every system that has just two choices is a binary system: On/off or True/False or Heads/Tails or Open/Closed would be examples of binary systems. I have summarized counting 0 through 7 in the following chart. Note the same pattern as the truth tables and the coin flipping...

Counting in base 10	Counting in base 2	Truth tables for 3 variables	Tossing three coins
0	0 0 0	T T T	H H H
1	0 0 1	T T F	H H T
2	0 1 0	T F T	H T H
3	0 1 1	T F F	H T T
4	1 0 0	F T T	T H H
5	1 0 1	F T F	T H T
6	1 1 0	F F T	T T H
7	1 1 1	F F F	T T T

Notice that the 0 in the **base 2** counting column corresponds to the **T** in the truth table chart and corresponds to the **H** in the coin-tossing chart. In like manner, the **1** in the **base 2** counting column corresponds to the **F** in the truth table chart and corresponds to the **T** in the coin-tossing chart! Basically, these three tables are **the same pattern**. It is the significance of the given pattern that is so powerful. These charts give us a logical and systematic method for creating all possible combinations of two things, be they in reference to counting, or truth tables or coins. The same table could be extended to counting higher, or more variables, or more coins. As (I hope) you will recall from chapter 2 of the Logic Project, the number of rows will always be 2^n. In this case it was 2^3, or 8 rows. If we were to count to 15 or have four variables or toss four coins, we would have 2^4 or 16 rows.

The final connection I am going to make is with the area of a square and the volume of a cube. If we draw a square that is A + B on a side, and find the area, it will consist of two squares and two rectangles. One square will have an area of A^2, each rectangle will have an area of AB, and the other square will have an area or B^2. Or, in other words, the total area would be $A^2 + 2AB + B^2$, a trinomial that I am sure you recognize! In like manner, a cube with a side of A+B will have a volume of $A^3 + 3A^2B + 3AB^2 + B^3$, another polynomial that I think you will recognize. A detailed description of these two problems can be found in the January 2000 issue of the *Mathematics Teacher*. I believe it would be well worth your time and energy to look this up! The implications of this technique go far beyond the scope of these projects and will be seen again if you continue your study of mathematics beyond the minimum required.

www.ingramcontent.com/pod-product-compliance
Lightning Source LLC
Chambersburg PA
CBHW051414200326
41520CB00023B/7229